For my beloved wild creatures,
Chase and Mackenzie
—and, of course, for Mom

Published by the National Geographic Society

ISBN: 978-1-4262-0922-2

The National Geographic Society is one of the world's largest nonprofit scientific and educational organizations. Founded in 1888 to "increase and diffuse geographic knowledge," the Society's mission is to inspire people to care about the planet. It reaches more than 400 million people worldwide each month through its official journal, *National Geographic,* and other magazines; National Geographic Channel; television documentaries; music; radio; films; books; DVDs; maps; exhibitions; live events; school publishing programs; interactive media; and merchandise. National Geographic has funded more than 9,600 scientific research, conservation and exploration projects and supports an education program promoting geographic literacy.

For more information, visit www.nationalgeographic.com.

National Geographic Society
1145 17th Street N.W.
Washington, D.C. 20036-4688 U.S.A.

For information about special discounts for bulk purchases, please contact National Geographic Books Special Sales: ngspecsales@ngs.org

For rights or permissions inquiries, please contact National Geographic Books Subsidiary Rights: ngbookrights@ngs.org

Interior design by Melissa Farris

Printed in the U.S.A.

12/WOR/3

mother's love

Canada lynx and cub

Hen and chick

mother's love

Inspiring True Stories from the Animal Kingdom

Melina Gerosa Bellows

FOREWORD BY KATE HUDSON

NATIONAL GEOGRAPHIC

WASHINGTON, D.C.

Mother and
infant chimpanzee

foreword

Becoming a mother changed so many aspects of my life immediately. From the moment I became pregnant with my first, Ryder, nature took over. That truth became clear to me one day when I took my son to see the new baby chimp at the London zoo. The chimp's mother was helping him climb a fence. She sat at the top and watched as each time he climbed a bit, he fell back down. Finally, close to the top, he lost his footing. Calmly, patiently, the mother reached her long hand down, pulled him up, and held him close. I started to cry. Her actions mirrored my own instinct: to let my kids feel trust and freedom without disrupting the journey. It's hard not to want to keep them away from any harm, but that's how we can teach our little "animals" to survive in the wild.

KATE HUDSON

actress

introduction

Am I going to lose my son?

The fear spears me like an ice pick as I stare at my six-year-old child lying on the floor, unable to move.

Until that moment, it had been a regular Sunday morning. "If you're not dressed and ready for church by the time I get there, there'll be trouble!" I called up to Chase and his five-year-old sister Mackenzie, both professional dawdlers.

Then I heard a loud thump, a few long seconds of silence, and the ensuing wail. Every mother knows that there are two kinds of crying: the usual kind, and the kind that makes your heart stop. This was the latter.

I run to Chase, horrified that I had dismissed his earlier claims of a stomachache as another subterfuge to skip Sunday school.

"I can't move my neck!" Chase is shrieking over and over, his squinty brown eyes drilling into me for help.

But I feel helpless, shackled by indecision. Chase's father

is out of town, so I'm on my own. Should I call 911? Or rush him to the nearby hospital? Would moving him intensify his injury? My swirling thoughts lock on a recent email we parents received, explaining that one of his classmates had come down with a mysterious life-threatening illness. Was it contagious?

My maternal instincts kick in and before I know it I'm in action, driving like a NASCAR racer to the ER. For the next two hours Chase is evaluated by three doctors and endures a battery of tests.

"Chase, close your eyes and count to ten," the doctor says.

He closes his eyes. We wait, but he doesn't say anything.

"Chase, can you count to ten for us?" she asks again.

"I am," he informs her. But still nothing.

"We can't hear you," she says. I panic, trying hard not to picture the possible diagnoses for a boy who thinks he is speaking but isn't.

Finally he speaks up. "I'm doing it in my head," he says, deadpan.

The doctor and I lock eyes and silently crack up.

Very quickly the scary stuff—broken neck, spinal meningitis, and even the Coxsackie viral infection his classmate had—is ruled out. It turns out Chase has a case of torticollis—the clinical name for a muscle spasm where you can't move your neck. He's going to be fine—and therefore, so will I. The reality of being a mother is you are only as well as your least well child. As author Elizabeth Stone said, "Making the decision to have a child is momentous. It is to decide forever to have your heart go walking around outside your body."

There may be no stronger power on Earth than a mother's love. Mothers have lifted cars, taken bullets, charged into burning buildings, and fought off bears to protect their children. But humans aren't the only ones who go to such elaborate lengths for the sake of their young—so do mothers in the animal kingdom.

There's Scarlett, the cat who ran into a burning building five times to rescue her kittens. There's the duck who convinced a strange human to drop everything and help her save her eight ducklings (how she knew a Good Samaritan when she saw one, we'll never know). And who can forget the dog that adopted a motherless baby squirrel and raised it along with her own puppies?

Working at the National Geographic Society for the last 12 years, I've been lucky enough to hear a wealth of such fascinating stories. If these reports from the field weren't from the world's best photographers and researchers, the tales would strain my credulity. Truth be told, however, I've become addicted to them. For some reason, the uncanny combination of "awe" and *"awww"* resonates with me.

Perhaps these stories touch me so deeply because they illustrate the common bond of motherhood. They show that mothers are mothers, no matter what species. Animals and humans alike have the natural ability to distinguish the unique cry of their own offspring from the cry

Mackenzie, Melina, and Chase Bellows

of other babies. We are hardwired to love, protect, and care for our young. After all, the survival of each of our species depends on it.

Still, science is reluctant to say that animals love in the same way that humans do. First, animals can't actually tell us how they feel, and second, scientists haven't yet been able to conduct conclusive experiments. Researchers can only extrapolate by judging an animal's facial expressions, vocalizations, body language, and behavior. There is, however, room for interpretation, as many scientists will agree.

Whether you choose to call that maternal bond love or instinct, these stories are nothing short of amazing. Each gives evidence of not only the instincts of motherhood but of other traits as well. Strength, tenacity, bravery, ingenuity, selflessness, perseverance, courage, resourcefulness—what human mother wouldn't aspire to qualities so admirable? Mother Nature has raised her daughters well.

These stories of mother's love within the animal kingdom greatly inspire me. I hope they will touch your heart, too.

A mother's arms
are made of tenderness and children
sleep soundly in them.

VICTOR HUGO

author

Born blind and deaf, kittens depend on their mother's sense of sight and hearing, navigating the world by touch, taste, and smell for the first three weeks of life.

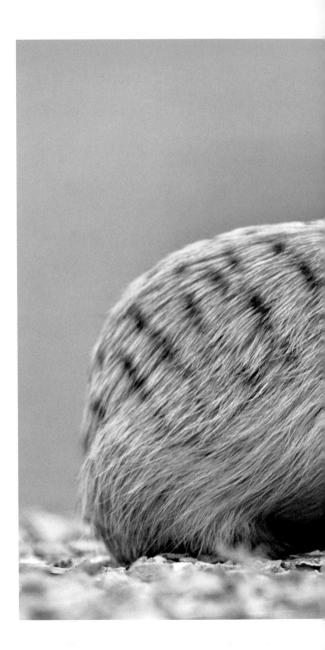

Sometimes
the strength of
motherhood is
greater than
natural laws.

BARBARA KINGSOLVER
author

In a meerkat mob, the top female gives
birth to most of the young, and the other
meerkat females help as "babysitters."

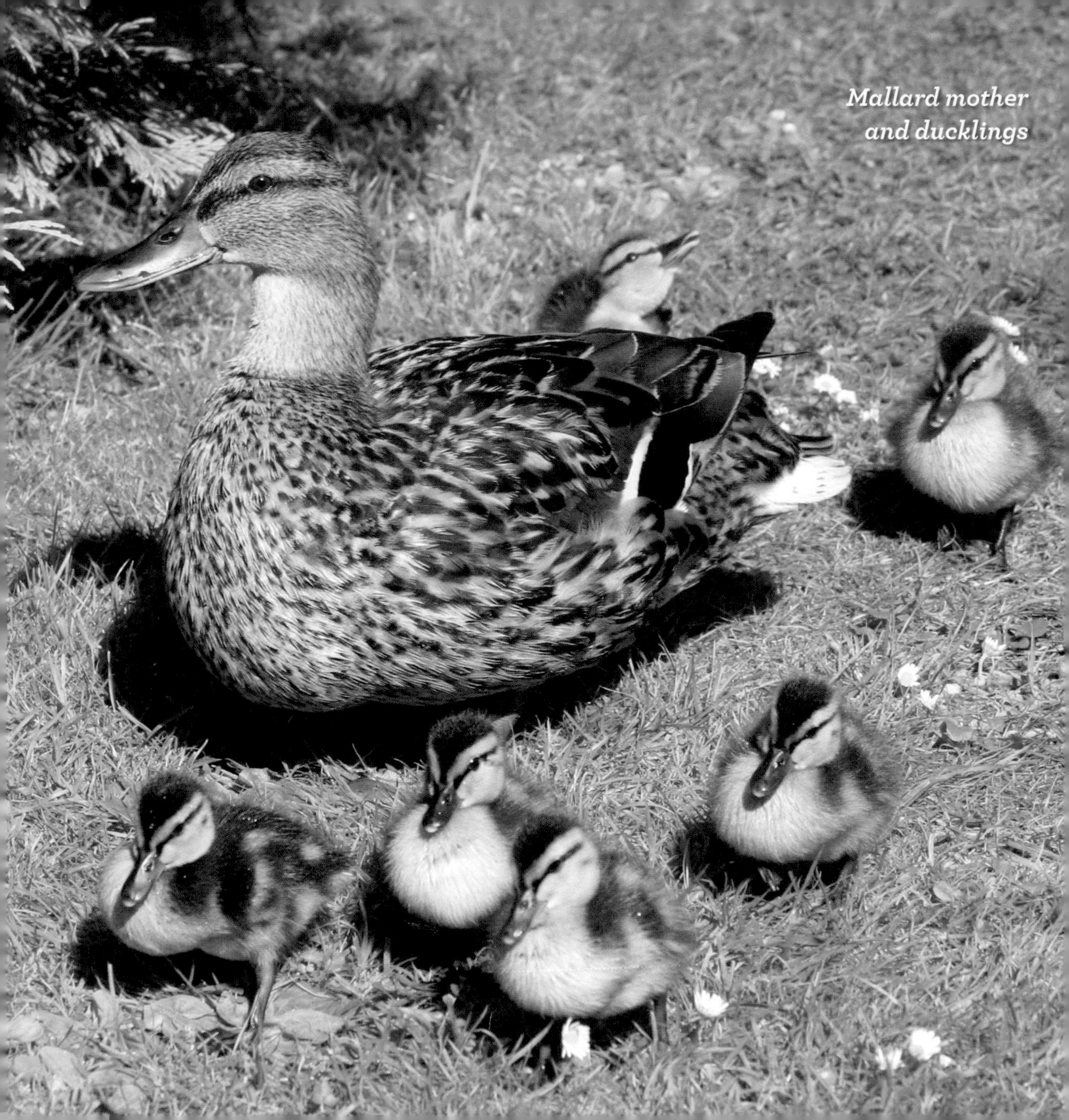

Mallard mother and ducklings

mom to the rescue

When a mother duck approached Vancouver resident Ray Petersen, flapping her wings and squawking like crazy, he didn't know what to think. She began running in circles around him and tugging at his pant leg with her beak. Finally, when she plopped down beside a sewer drain, he got the message. Looking inside, he saw that her eight ducklings had fallen through the grate to the sewer below—and Petersen knew he had to launch a rescue mission. He called the police and a tow truck to remove the 200-pound grate cover. Then, using a vegetable strainer, he lifted the ducklings, one by one, out of the sewer to safety. Once the mother and babies were safely reunited, the happy family waddled away.

Some hippos are born underwater.
When that happens, the mother nudges
her newborn up to the water's surface
so it can breathe.

My mother was the most splendid woman I ever knew... If I have amounted to anything, it will be due to her.

CHARLES CHAPLIN

actor

A lamb may use sight, sound, or smell to single out its mother from all the other ewes.

Making the decision to have a child
is momentous. It is to decide forever
to have your heart go walking
around outside your body.

ELIZABETH STONE
author

*A mother spectacled langur—a Malaysian monkey—
nurses her baby for a year.*

a mother's strength

As a mother elephant and her calf waded peacefully in Myanmar's Taungdwin River, a flash flood engulfed them. Although the mom—a log-hauling elephant named Ma Shwe by workers at a nearby camp—clung to her baby, the floodwaters separated them. Ma Shwe struggled to regain hold of her calf. She hoisted the baby up with her trunk, stood on her hind legs, and lifted the little one to safety on a high ledge above the river. Exhausted, the mother elephant fell back into the swirling waters and was swept away herself. Minutes later, a camp manager who had witnessed the rescue heard what he later called "the grandest sounds of a mother's love I can remember"—majestic elephant trumpeting. Calling loudly to her baby, Ma Shwe had emerged downstream and was tearing back through the forest for a reunion with her child.

Indian elephant mother and calf

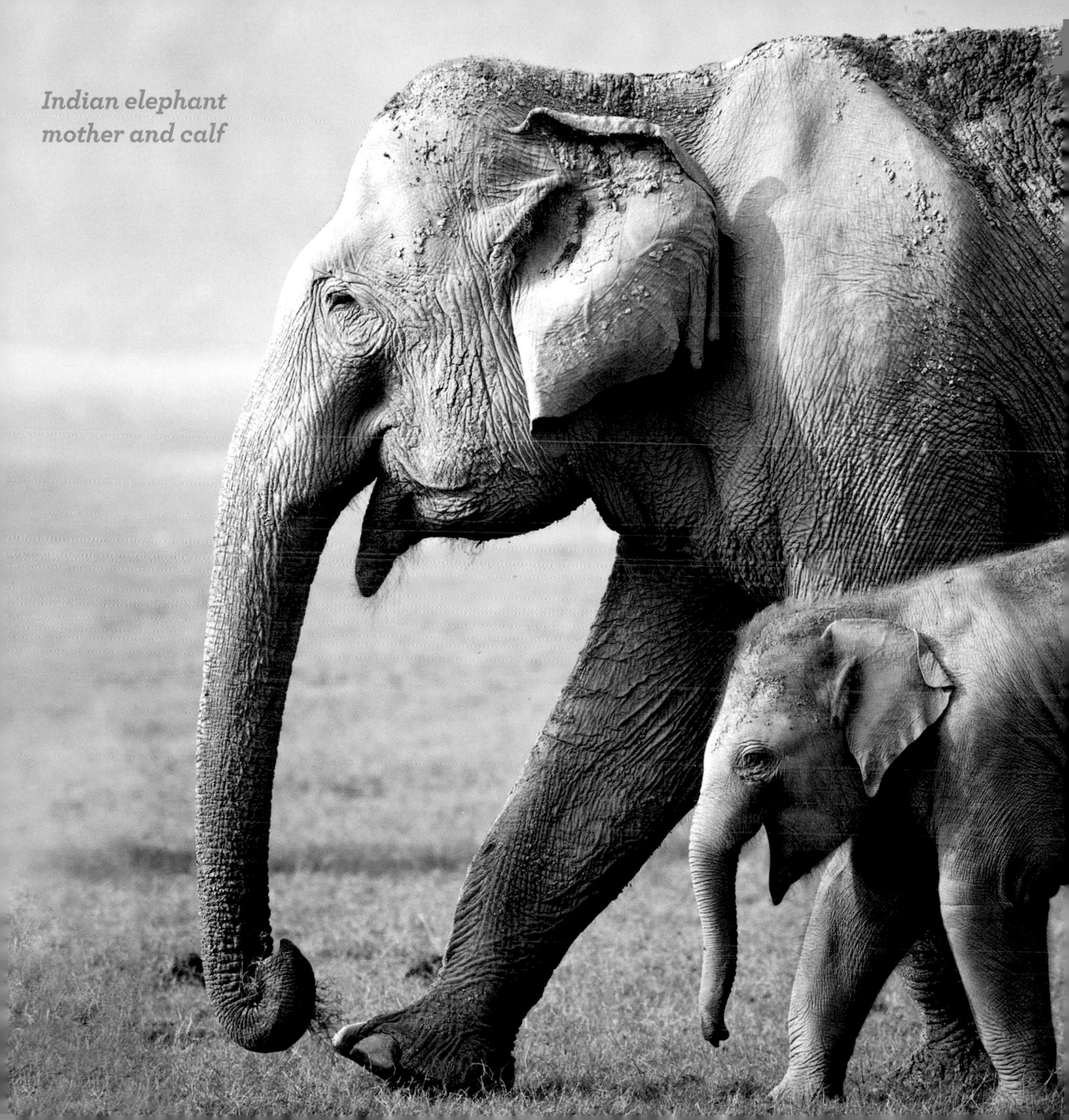

Mother is the home
we come from,
she is nature, soil,
the ocean.

ERICH FROMM
psychologist

*A Canada goose sits on her eggs for a month,
turning and repositioning them often to be
sure she warms all equally.*

My mother was the making of me.
She was so true, so sure of me,
and I felt I had someone to live for,
someone I must not disappoint.

THOMAS ALVA EDISON

inventor

At birth, a calf needs its mother's touch to get blood circulation going.

Beluga whales weigh nearly 200 pounds
at birth and nurse for up to two years.

mother mädchen

Mothering instincts can create unlikely families, like the baby deer and the dog who adopted her. When animal control officers in Wiesbaden, Germany, brought an orphaned fawn to the home of animal rescuer Lydia Weber, she expected to care for the baby animal herself. Then her dog Mädchen stepped in. Mädchen, a German shepherd mix, took to the orphan immediately and began licking and grooming her from head to toe. When Weber tried to bottle-feed the deer, now named Mausi, the fawn resisted drinking the milk—until Mädchen came to the rescue again. The dog lay down and patiently allowed Weber to place a bottle snugly against her belly, as if the dog were nursing the fawn. Comforted, Mausi hungrily suckled away, tended by her new mother, Mädchen.

Mädchen and Mausi

A mother's love for her child is like nothing else in the world. It knows no law, no pity, it dares all things and crushes down remorselessly all that stands in its path.

AGATHA CHRISTIE
author

Mother cheetahs use a birdlike "chirrup" sound to call their cubs.

I cannot forget
my mother.
She is my bridge.
When I needed
to get across, she
steadied herself
long enough for me
to run across safely.

RENITA WEEMS

minister

*Newborn sea otters can float, but they can't
swim. Mothers balance them on their stomachs.*

Born the size of honeybees, opossums live in their mother's pouch for two months before emerging to see the world.

mother of invention

Japanese macaques can convey ideas to one another and pass skills down, generation to generation. In 1953, Japanese primate researchers observed an alpha female monkey they named Imo dip a dirty sweet potato into the water to clean the sand from its skin before eating it. She also developed an ingenious method of cleaning sand from wheat by tossing the sandy grain into water. The sand sank, but the lightweight grain floated, and she was able to pick it out and eat it. Soon, other members of her troupe caught on, and before long they had adopted both practices. Imo's generation passed the skill to their offspring, and monkeys in this area of Japan still use it today.

Japanese macaque
baby and mother

The phrase "working mother" is redundant.

JANE SELLMAN
author

One garden snail can have as many as 430 babies in one year.
The female lays between 30 and 120 eggs at a time.

There are a
million moving parts
to raising kids, and
you can't always
anticipate them all...
It's as simple, and
as scary, as that.

ANNA QUINDLEN
author

*A mare chooses the company her newborn
keeps. Older siblings and trusted humans may
approach, but she wards off all others.*

my mom, my hero

While gathering food on the banks of Malaysia's Segama River, a baby orangutan and mother were stranded up a tree by the rushing waters of a flash flood. When conservation groups arrived on the scene, they had to think fast: Interacting with a wild, frightened, and protective orangutan mother could be dangerous. They decided to tie a rope to a tree and toss her the free end. With her baby clinging to her back, the mother grasped the rope and half-pulled, half-swam to the safety of the jungle shore. "We rarely see orangutans in the water," says Serge Wich of Iowa's Great Ape Trust. "But these moms go to great lengths to save their babies."

Orangutan mother and infant

Mother foxes use play
to teach their kits
fighting and hunting skills.

My mother's menu consisted of
two choices: take it or leave it.

BUDDY HACKETT

comedian

Mother sunbirds build their nests and hatch their eggs on their own,
but the fathers return to the nest when it is time to help feed the hatchlings.

Giselle and her puppies with Finnegan the squirrel

room for one more

Mademoiselle Giselle, a dog owned by wildlife rescuer Debby Cantlon, doesn't waste her time chasing
squirrels—she adopts them. A three-month-old squirrel had been found abandoned beneath a tree and
was brought to Cantlon's home. Giselle, pregnant
herself at the time, adopted the baby squirrel, now
named Finnegan, and maneuvered his cage next to
her dog bed. Cantlon tried to move Finnegan back
to his own corner, but Giselle persisted and again
brought the squirrel cage to her side. With some
trepidation, Cantlon released Finnegan into Giselle's
bed—and the pair bonded. Giselle soon gave birth to
five Papillon puppies, which she licked and nursed
right alongside their squirrel brother. Finnegan even-
tually returned to the wild, but he does return to the
Cantlon house to visit.

Children are the anchors that hold a mother to life.

SOPHOCLES
playwright

Baby ring-tailed lemurs rarely venture more than two feet away from their mothers.

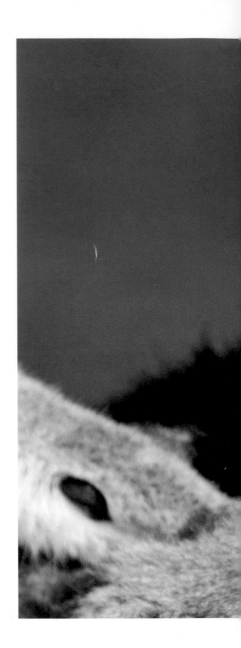

A polar bear is capable of giving birth
and nursing her newborn cubs
during her long winter sleep.

A mother is not a person to lean on but a person to make leaning unnecessary.

DOROTHY CANFIELD FISHER
educator

After nurturing her cubs for about two years, a mother lion pushes the males out on their own, but she and her daughters may remain together for life.

a family affair

Marine mammal specialist Richard O'Barry came upon a baby dolphin who had become entangled in a nylon fishing net. He watched with horror as the dolphin's mother and two other females stopped at nothing to save him. "They tried to push that baby out—at the risk of tangling up their own fins," remembers O'Barry, who ultimately dove in and cut the net himself. The mother then guided her weakened, gasping calf up to the water's surface and supported him so he could breathe. Exhausted by the ordeal, she, too, struggled to float, and the other females supported both mother and child, helping them regain their strength. For the next few hours, all three adults took turns keeping the baby afloat until finally, to O'Barry's relief, the entire extended family swam away.

Bottlenose dolphins,
mother and calf

My mother
had a great deal
of trouble with me,
but I think she
enjoyed it.

MARK TWAIN
author

Mother dogs lick their newborn pups clean,
biting off each umbilical cord, an essential step
in the bonding process.

Porcupines are covered with 30,000 barbed quills, but a porcupette's are soft and bendable at first, hardening the first few days after birth.

A mother always has to think twice,
once for herself and
once for her child.

SOPHIA LOREN

actress

*A giant panda might nurse her young as often as 14 times a day,
30 minutes at a time.*

Mother grizzly
bear and cubs

don't mess with momma

In Alaska's Katmai National Park, a mother grizzly and her two cubs came to the water's edge for a meal. The two cubs, nicknamed Sugar and Spice by wildlife photographer Jim Abernathy, played while mom fished for salmon about 150 yards away. Suddenly a huge male grizzly emerged from the trees, lumbering toward the cubs. Lone males can be deadly, and the cubs sensed it. Panicking, Spice sprinted toward his mother, but the faster adult male caught him. Things looked grim until a blood-curdling momma bear's roar rang out. Mother bear charged the larger, startled grizzly. He flung Spice aside and ran, chased mercilessly by the mother grizzly. "She clearly wanted a piece of him," Abernathy remembers. Mother bear gathered her cubs, made sure they were safe, and then returned to her fishing.

To describe my mother would be to write about a hurricane in its perfect power. Or the climbing, falling colors of a rainbow.

MAYA ANGELOU

poet

A leopard leaves her cubs for up to 36 hours to hunt; on her return she is likely to move them to a new lair for safety.

A child of one can be taught not to do certain things, such as touch a hot stove, turn on the gas, pull lamps off their tables by their cords, or wake Mommy before noon.

JOAN RIVERS
comedienne

When food is scarce, great gray owl mothers will starve themselves to feed their owlets. Some lose up to a third of their body weight.

A mother harp seal can instantly recognize her pup using only her sense of smell.

It is not until you become a mother that your judgment slowly turns to compassion and understanding.

ERMA BOMBECK
author

After a 12-month gestation period, donkeys are born fully developed. They stand on their own and nurse within 30 minutes of birth.

Scarlett the cat

the bravest of them all

Scarlett the cat never wavered when danger threatened her month-old kittens. As a four-alarm fire raged in the abandoned Brooklyn garage that was their home, Scarlett ran back and forth through the burning building five times to retrieve every one of her kittens. Fur singed, ears scorched, and eyes blistered from the intense heat, Scarlett arranged her kittens and touched each one with her nose to make sure they were all there. Then the badly injured mother cat collapsed. Firefighter David Giannelli brought the family to the North Shore Animal League, where the mother and kittens were treated for their injuries. As Scarlett recovered, the story of her bravery spread. By the time she was healthy again, more than 7,000 people had offered to adopt her. She and her kittens found caring homes.

Whatever else is unsure in this
stinking dunghill of a world
a mother's love is not.

JAMES JOYCE
author

*A dominant mother pig typically gives birth to more male piglets
than the subordinate sows.*

Children...require guidance and sympathy far more than instruction.

ANNE SULLIVAN
Helen Keller's teacher

Giraffe mothers give birth on their feet—and their newborns drop six feet to the ground, headfirst. The fall actually helps them take first breaths.

As soon as goslings hatch, they start to follow the first moving object they see, establishing a lasting connection. Most of the time that moving object is the mother goose.

Sita the Bengal
tigress and
her cub